AGILE PROJECT MANAGEMENT FOR BEGINNERS

The most complete Guide To Scrum, Agile Project
Management, Agile Software Development and Project
Management

GEREMY WILSON

not engaging in the rendering of legal, financial, medical or professional advice. The content within this book has been derived from various sources. Please consult a licensed professional before attempting any techniques outlined in this book.

By reading this document, the reader agrees that under no circumstances is the author responsible for any losses, direct or indirect, which are incurred as a result of the use of information contained within this document, including, but not limited to, errors, omissions, or inaccuracies.

Table of Contents

INTRODUCTION

Agile in English is the ability to move and respond quickly. In software development, Agile constitutes various ways of developing software through the collaboration of team members and end-users. The goal of Agile software development is to deliver value faster, improve quality, allow for change and focus on end-users.

The Agile manifesto, which is a published declaration of the views of seventeen leading software developers in the software industry composes of shared experiences in regards to what works and does not work in software development. The developers were looking for solutions to the current software development problems like unresponsiveness, too much documentation, and cumbersomeness. Therefore, they came up with twelve Agile software development principles and four values.

The Agile manifesto was not a declaration of facts but a comparison of different approaches, which is why many people embraced it. Under the manifesto, people received the titles of valuable resources other than just members of the workforce. Previous methods relied more on resources than on people. They also made it possible for people to go back and forth on their projects, get feedback from end-users, make changes, and continuously test their work. With the traditional methods, developers had to come up with a plan upfront and complete every stage before proceeding.

The manifesto also allowed for more flexibility where developers could ship software in time as opposed to shipping when the work was already obsolete. Unlike the previous methods that encouraged developers to build software to completion, the Agile development emphasized developing software in various stages.

Development Of The Agile Manifesto

The Agile manifesto leads to the creation of software, which offered its users new versions or releases after short periods of work called sprints. This approach was different from the traditional methods where the developers had to compile all their work first and then building the software all at once. When using the traditional method, it was only after completion that the people involved released the project.

Traditional methods expected developers to come up with the software layout and design upfront. Then actual implementation followed the direction of the design. People considered a project successful when development went according to plan. With Agile, they still make plans, which they revise constantly. Here the value delivered by the software is the basis on which the project is successful.

Sharing the project in sprints also has its advantages. If the people concerned detect a problem in the early stages, it is cheaper and easier to fix it than if one has to wait and fix it after completion.

What is Agile Project Management, Exactly?

This might be the first question that will pop up in your head and it is best that we answer it first. The most basic definition of APM is that it is a methodology in which work is completed and delivered in short cycles (sometimes called sprints) and tweaked over time. This is done with the goal of providing value through increasing performance results as well as improving the structure of the project plan. This has similar features as most project planning and management methods in which the entire project is divided into smaller tasks.

The goal of the methodology is to simply cut out the cost by doing away with the trial-and-error strategy used by many project managers. By using a more cautious and yet proactive approach to the project, one should be able to get better results from it.

Sounds straightforward, right? If you are still wondering what the style is, it would be better if we compare it to other methodologies.

The Flaws of the Classic Method

The classic project management method is simply the methodology that every manager typically uses as a default. It can differ from one manager to another, but it mostly manifests the same flaws in each project. They are the following:

A. Too Much Time on the Planning Phase

A major flaw with the classic project management method is that you are planning the entire project upfront. This means that you would spend more time finding the resources, evaluating costs, and assigning tasks and schedules for your team to follow.

The problem here is that a major portion of the time you spent for planning your strategy can be used more productively, like actually executing the plans. And the long time spent on planning also leaves the method open for another flaw, which is... etc.

B. Plans Become Inflexible Eventually

When you spend too much time on planning, you become dead-set on implementing it no matter what. A lot of managers do this because they don't want to waste all that mental energy they have spent creating the plan.

However, this does mean that the plan does not account for changes in the middle of the process. There will always be a chance that new deadlines have to be met which, in turn, requires the acquisition of new resources, the reshuffling of teams, and evaluation of new costs. This means that you can easily shoot over the budget and make the development process even more chaotic than necessary.

C. Treating the Team as Mere Resources

During planning, miscalculations can occur. Tasks or even entire projects could get underestimated. This would usually result in changes in schedule and resource allocations.

If you are not careful in this phase, you are inadvertently giving the impression that your team is expendable. Worse, you could just attribute failure in the implementation of the changes over the team. This can increase animosity between teams and team members that ultimately impedes everybody from achieving shared goals.

What Makes Agile Project Management Different

More often than not, it is preferred that you use any other methodology instead of the standard method. This is where Agile Project Management comes into play and it has a few characteristics that instantly set it apart from the standard method.

1. It is Segmented by Design

The agile methodology is comprised of several "iterations" that last about four weeks. This can help managers in the sense that it makes evaluation all the easier to perform on a regular basis.

Also, the subdividing of the larger project into several iterations can psychologically help managers and team members focus on developing specific parts of the project. If one part of the project has to be done in a later iteration, then there is no need to worry about completing its sub-goals; just yet.

2. It is Based on Time Periods

In as much as rojects in the agile method are divided into segments, each segment itself has a fixed time period. As was stated, each iteration can last up to a month, and this cannot be modified down the line if the team agrees to such conditions.

With their efforts focused on one part of the project for a short period of time, the agile method can help teams become more productive and generate observable results within each iteration. And this is even if there are changes brought about by outside forces during each cycle.

3. It is Easy to Understand, even for Non-Tech Folk

One of the problem areas in project management is actually in communicating progress to clients, shareholders, and other people who might not have a good grasp on the more technical aspects of your work. In some methodologies, progress is either rarely communicated on a regular basis as they are long-term in nature. Basically, there is nothing to report regularly as overall goals take a while to be achieved.

And if some goals are short term, they are communicated in a manner that is highly technical and confusing. All the clients see are the pretty graphs and numbers. Nothing about it seems to inspire their minds or give them the assurance that everything is running according to the plan.

Since the agile method runs on iterations, clients and the powers-that-be in your company can be certain of regular reports regarding the progress of a project. And not only will the reports be regular, they are presented in a manner that is easy to understand. This should give the clients an impression of what is going on and what to expect next.

The central idea behind Lean Analytics is to enable a business to track and then optimize the metric that will matter the most to their initiative, project, or current product.

Setting the goal of focusing on the right method will help you see real results. Just because your business has the ability and the tools to track many things at once, does not mean that it would be in your best interest to do so.

Tracking several types of data simultaneously can be a great waste of energy and resources and may distract you from the actual problems. Instead, you will want to focus your energy on determining that one vital metric. This metric will make the difference in the product or service that you provide.

The method in your search for this metric will vary depending on your field of business and several other factors. The way that you'll find this metric is through an in-depth understanding of two factors:

- The business or the project on which you're presently working.

- The stage of innovation that you are currently in.

Now that we have a basic understanding of Lean Analytics and what it means, let's take some time to further explore and see its different parts.

Lean is a method that is used to help improve a process or a product on a continuous basis. This works to eliminate the waste of energy and resources in all your endeavors. It is based on the idea of constant respect for people and your customers, as well as the goal of continuously working on incremental improvements to better your business.

Lean is a methodology that is vast and covers many aspects of business. This guidebook will spend sufficient time discussing a specific part of Lean, Lean Analytics. Here, you can learn how to make the right changes. Of course, you will need a working understanding of where to start, and Lean Analytics can help.

Lean is a method that was originally implemented for manufacturing. The idea was to try to eliminate wastes of all kinds in a business, allowing them to provide great customer service and a great product while increasing profits at the same time. Despite its beginnings, the Lean methodology has expanded to work in almost any kind of business. As long as you provide a product or a service to a customer, you can use the Lean methodology to help improve efficiency and profits.

Lean Analytics

Lean Analytics is part of the methodology for a lean startup, and it consists of three elements: building, measuring, and learning. These elements are going to form up a Lean Analytics Cycle of product development, which will quickly build up to an MVP, or Minimum Viable Product. When done properly, it can help you to make smart decisions provided you use the measurements that are accurate with Lean Analytics.

Remember, Lean Analytics is just a part of the Lean startup methodology. Thus, it will only cover a part of the entire Lean methodology. Specifically, Lean Analytics will focus on the part of the cycle that discusses measurements and learning.

It is never a good idea to just jump in and hope that things turn out well for you. The Lean methodology is all about experimenting and finding out exactly what your customers want. This helps you to feel confident that you are providing your customers with a product you know they want. Lean Analytics is an important step to ensuring that you get all the information you need to make these important decisions.

16

Before your company decides to apply this methodology, you must clearly know what you need to track, why you are tracking it, and the techniques you are using to track it.

Focus on the fundamentals

There are several principles of Lean that you will need to focus on when you work with Lean Analytics. These include:

- A strive for perfection

- A system for pull through

- Maintain the flow of the business

- Work to improve the value stream by purging all types of waste

- Respect and engage the people or the customers

- Focus on delivering as much value to the customer as effectively as possible

Waste and the Lean System

One of the most significant things that you will be addressing with Lean Analytics, or with any of the other parts of the Lean methodology, is waste. Waste is going to cost a company time and money and often frustrates the customer in the process. Whether it is because of product construction, defects, overproduction, or poor customer service, it ends up harming the company's bottom line.

There are several different types of waste that you will address when working with the Lean system. The most common types that you will encounter with your Lean Analytics include:

- **Logistics:** Take a look at the way the business handles the transportation of the service or product. You can see if there is an unnecessary movement of information, materials, or parts in the different sections of the process. These unnecessary steps and movements can end up costing your business a lot of money, especially if they are repeated on a regular basis. This will help you see if more efficient methods exist.

- **Waiting:** Are facilities, systems, parts, or people idle? Do people spend much of their time without tasks despite the availability of work or do facilities stay empty? Inefficient conditions can cost the business a lot of money while each part waits for the work cycle to finish. You want to make sure that your workers are taking the optimal steps to get the work done, without having to waste time and energy.

- **Overproduction:** Here, you'll need to take a look at customer demand and determine whether production matches this demand or is in excess. Check if the creation of the product is faster or in a larger quantity than the customer's demand. Any time that you make more products than the customer needs, you are going to run into trouble with spending too much on those products. As a business, you need to learn what your customer wants and needs, so you make just the amount that you can sell.

- **Defects:** Determine the parts of the process that may result in an unacceptable product or service for the customer. If defects do exist, decide whether you should refocus to ensure that money is not lost.

- **Inventory:** Take a look at the entire inventory, including both finished and unfinished products. Check for any pending work, raw materials, or finished goods that are not being used and do not have value to them.

- **Movement:** You can also look to see if there is any wasted movement, particularly with goods, equipment, people, and materials. If there is, can you find ways to reduce this waste to help save money?

- **Extra processing:** Look into any existing extra work, and how much is performed beyond the standard that is required by the customer. Extra processing can ensure that you are not putting in any more time and money than what is needed.

How Lean can help you define and then improve a value stream

Any time that you look at the value stream, you will see all the information, people, materials, and activities that need to flow and cooperate to provide value to your customers. You need these to come together well so that the customer gets the value they expect, and at the time and way, they want it. Identifying the value stream will be possible by using a value stream map.

You can improve your value stream with the Plan-Do-Check-Act process.

Another method of creating this environment is the 5S+ (Five S plus): sort, straighten, scrub, systematize, and standardize. Afterward, ensure that any unsafe conditions along the way are eliminated.

The reason that you will want to do the sorting and cleaning is to make it easier to detect any waste. When everything is a mess, and everyone is having trouble figuring out what goes where, sorting and cleaning can address waste quite fast. There will also be times when you deem something as waste and then find out that it is actually important.

When everything is straightened out, you can make more sense of the processes in front of you. Afterward, you can take some time to look deeper into the system and eliminate anything that might be considered as waste or unsafe, and spend your time and money on parts of the process that actually provide value for your customer.

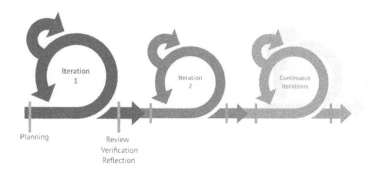

Planning

Review
Verification
Reflection

Iteration 1

Iteration 2

Continuous iterations

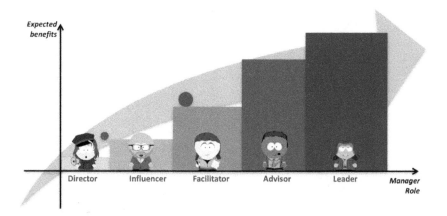

Expected benefits

Director   Influencer   Facilitator   Advisor   Leader

Manager Role

Chapter 1: Drawbacks of Agile

Weaknesses To Look Out For In Agile

Has Agile helped in software development in a way that is satisfying to the end-user? Yes. Have the principles of Agile project management helped to raise efficiencies and profitability in other sectors like manufacturing and construction? Definitely yes!

Here is a list of Agile weaknesses that serious leaders in the field of IT as well as seasoned business people are checking out as they prepare to adopt Agile in their organizations:

Many stakeholders do not understand Agile development

There are those people that confuse the use of the Agile development approach with the agility denoting flexibility with abandon; or with no controls whatsoever. Surely, if that is the meaning your team will have as they implement your Agile projects, then you are very likely to fail in pleasing your consumer or end-user of your product.

Challenges in interacting with stakeholders

So Agile encourages – in fact, demands – that the team receives input and feedback from the product owner and other stakeholders on a regular basis. Who is to assure their availability? Surely your team or even your CEO cannot reassure you of their consistent availability.

Agile calls for co-teams to be housed together

This is to say that an Agile project is sometimes carried out by some small teams that ultimately make one big Agile team for the project. And because the nature of Agile Project Management is one of cooperation and coordination, such teams need to be in close proximity to one another.

Hitches in scaling projects

Evaluations and assessments of projects and various iterations before and during the process are crucial in Agile developments. That is often an easy task when the team is small and manageable; say 3 – 8 people as is the norm. Just to recap the composition of an Agile team, we have: the product manager, the team leader, one or even up to two developers, and sometimes you have a designer, a business analyst, as well as a tester.

Challenges in Architectural Planning

Sometimes the Agile team lacks this as they begin the project. Yet, just as in the building architecture, software architecture is required up-front; that is before any work on software development can begin. In the building world, you would not know how well to utilize materials and integrate your structure into the landscape in the absence of an architectural plan. Likewise in software development, if you have no idea what the platform to be used is, and possibly your architectural approach is pretty new and untested, it becomes tricky.

Limited planning in Agile

Prior planning in projects is always helpful. It even helps to analyze eventualities and prepare for them. However, although Agile has planning tools, in practice, little emphasis is put on planning (preference being given to use of backlogs, instead). Then iterations become the guide for daily performance where the aim is to complete product features. This is as opposed to delving much into the entire scope of the project. This form of lightweight planning where estimations and tracking are based on small work bits is only suitable for small enterprises where you are mostly dealing directly with the product owner.

Need for re-work

Sometimes in Agile development, different components of a product need to be integrated to make the whole product work. That kind of interaction is not easy in a methodology that does not embrace long-term planning, and one whose importance on advance architecture is light weight. Of course it is known in Agile projects that at some point different parts will need to be combined, even when you are talking of software development, and this is known as refactoring. The cost of refactoring is even incorporated in the project budget.

Obstacles in committing to contracts

Whenever a project is to be undertaken, it is important to have a comprehensive contract at hand before the work begins. This contract includes the expected expenditure or even the predictable cost estimates. In Agile development where projects are priced per iteration and a lot of leeway is given for changes even after the project has begun, it becomes difficult to provide a reliable contract amount, particularly for contracts that of the nature of fixed bids. This is not just frustrating to the client, but also to senior management who are better placed to give a nod or a decline when conclusive figures are involved.

Agile inhibits continuity

As has already been stated, the Agile development process undervalues documentation putting more prominence to personal interactions between project team members and the clients. Even when it comes to developing work segments within iterations, nobody cares much about written material as long as the client likes the product. What then happens when one of the project team members leaves the team, possibly to another organization? It is even worse if the whole team departs with all its knowledge about the development of the software or whatever other product they worked on. It would mean that if there was the need to upgrade the product, the organization would be at a great loss. That weakness emanates from letting the knowledge pertaining to the project remain just with the individuals involved with the Agile development process.

Challenge in integrating with other systems

The lack of extensive and detailed prior planning is a challenge in Agile development. Often the end-user will require some input from other systems like the accounting department or even the order processing personnel. It would help the application greatly if the project provided for the appropriate integration points in advance. Other times it could be the software requiring integration points in order to connect with the system monitoring inventory as well as order processing.

The Weaknesses of Agile

As it typically occurs, the greatest strength is also its greatest weakness. Flexibility can result in a lack of attention and motivation to complete your project if you do not watch over it. Having a loose plan instead of milestones means there is no set process to check in on and see that there is a smooth progression. This looseness can result in the team losing focus. To combat this weakness, consider creating an internal process to run alongside Agile to help keep your teams on target, or consistently check in to ensure your teams are constantly communicating and moving onward. Sometimes you may even need to consider one of the offshoots of Agile if you continue to find this weakness tripping up your teams.

How Agile Can Easily Fail

It is also crucial for you to learn a thing or two about how you can fail when using the agile approach. There is no agile technique where the goal is a failure. However, poor planning by an informed team can easily lead to defeat. You might be using the best agile method, but it will not work if you do not use the right approach. The following lines will discuss how things can go wrong with the agile strategy. This information will provide you with the insight you need to avoid common pitfalls, which could hinder your organization from succeeding.

Planning Chaotically

A common myth surrounding the use of the agile method is that planning is not important. This is not true. In spite of having an agile system in place, it doesn't mean that planning is not important. Planning is the best way to execute your strategy. Fortunately, there is a tool to help your team plan effectively align their actions to the business goals. This tool is called the Agile Release Train, and it can help a company to plan projects even twelve months in advance.

Forming an Unstable Team

A lot has been said about the collaboration. The success of any agile program depends on how well your team can combine their efforts. If you work with a team of individuals who understand each other, then you will have the upper hand. On the contrary, if your team is made up of strangers, it will be daunting for you to achieve the success you are looking to achieve.

Teams should be optimized to aid in reducing dependencies. An ideal group must be comprised of individuals who understand their job specifications. They should not wait for the project manager to guide them constantly. The best agile teams are made up of generalists. These are experts who will not depend on other specialty groups to complete their tasks.

Communicating Infrequently

Agile teams with poor communication will fail quickly. The flow of communication in an agile environment should be continual. Regular interaction should be maintained with stakeholders who have an interest in the end product. The project manager should work to make sure that all members of the organization converse on similar wavelengths. It is quite likely that there will be certain terminologies that will be used in the workplace. A good team should have a uniform vocabulary concerning the product.

Poor Testing of Products

Having regular tests guarantees that quality is maintained when developing products. Poor testing is the fast road to failure. It is crucial for developers to accurately test products to make sure that they are up to standards. Regular tests aid in guaranteeing that the team finds it comfortable to adjust to changes.

Failing to Comprehend the Project's Scope

To ensure that the agile management approach functions optimally, it is crucial for the project manager to create a roadmap. This roadmap guides the development team on how tasks will be carried out. It lays the path to be taken to ascertain that a functional product is obtained at the end of each sprint. Therefore, it is important that all members of the team understand the project's scope from the start. This ensures that every step is in line with the project's goals.

Disregarding Customer Feedback

Customer feedback is what will bring about changes in the agile system. Failure to heed to consumer evaluation only leads to frustration. Your team will be disappointed that in spite of their efforts, they did not succeed. Customer comments should be highly regarded as they warrant that the product's requirements are fully met. The absence of customer feedback makes it a daunting task to know how to prioritize. Ultimately, the development team might end up disregarding what is most important in the eyes of the customer.

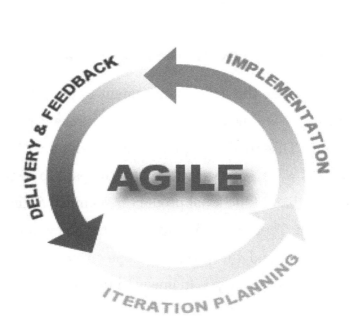

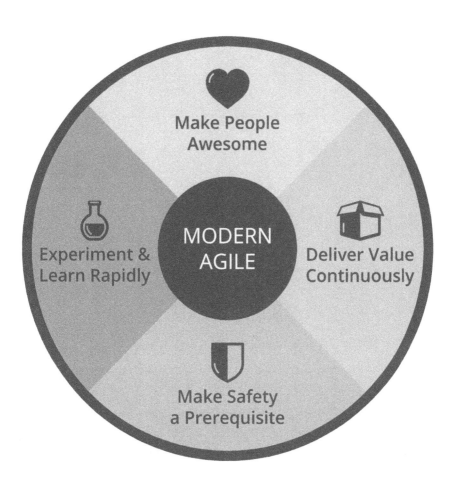

Chapter 2: Software, Tools and Techniques

The Agile Software Development Process

There is actually no single methodology out there that can work for every project. However, there is no doubt that many development teams and companies are slowly doing away with the more predictive and restrictive methodologies like Waterfall and embracing something more adaptive like Agile.

How Development was Done Before

So what was the development process in methodologies like, say, Waterfall? However, it is still necessary that we make a distinction over predictive methodologies over adaptive ones.

The conventional software development process will involve six phases which are as follows:

Planning

Obviously, every development process starts with you laying down the specifications of the project. Here, the flow of work will be identified and segmented into smaller and more manageable parts.

The functionalities of each segment and element will also be identified as well as the schedule for each phase of the project. Lastly, the workload will be identified here as well as the roles that each member of the development team would perform.

Analysis

This part will involve the identifying of goals as well as setting the scope for the entire project. This is a far more detailed process than the planning phase, as each stage of the project will be scrutinized.

A major focus of this phase is identifying the allocation of resources for each part of the project. What is the budget for each phase? What are the tools and programs needed? Is there a need to outsource work or hire entirely new people for the job (even temporarily)? These questions need to be sufficiently answered at this part of the process.

Of course, this process will also involve identifying potential issues that might pop up in the middle of the project. In turn, this allows managers to come up with solutions to prevent such from happening.

Design

Once planning and analysis have been completed, the team can move on to designing the product. This is a purely conceptual phase as you and your team would visualize what the project looks like by setting up its framework.

Here, the standards for each phase of the project will be established. As such, the team knows what they have to do in order to produce the desired software while also eliminating flaws.

Development and Implementation

This is the phase where the product is actually being built. Depending on the chosen methodology, this phase will involve multiple processes that include code writing and the implementation of programming tools and languages.

Once the software is developed, the implementation process kicks in where it goes through various studies and experimentations to see if it, at the very least, functions without crashing.

Testing

Once the basic structure of the software is finished, it will then go through a series of tests. Here, the goal is to identify bugs and glitches embedded into the code through the development process and then to fix them.

Just like the development process, this is a rather extensive phase as the program has to be scrutinized in all of its aspects and functions to see if it is fit for mass production and distribution.

The most important aspect to be tackled here is determining whether or not the product meets the criteria set in the initial phases of the project. In some cases, the overall layout of your program would be changed in order to address inherent flaws.

Maintenance

Prior to mass production, the team should then systematically scour the code for any bugs or glitches that were not identified and addressed in the previous phases.

This part also includes updates that would be introduced way after the product has been released.

Flaws in the Conventional Method

Almost all predictive methodologies follow the sequence as laid out above. However, some methods like Waterfall would like to add a few more steps in between, such as Research and Feedback.

Whatever the case, predictive methodologies tend to follow a strict sequence in order to create a product that works. However, that does not mean that it is applicable in all cases.

As a matter of fact, there are flaws inherent to these methodologies which may make them inapplicable to your project or, better yet, inferior to other more adaptive methodologies.

Restrictive Nature

At a glance, predictive methodologies are so rigid that you have no other option but to follow the plan as was established in the earlier phases. Of course, this means that you are not exactly responsive to changes as they occur in the middle of the project.

In the end, you will produce something that might meet the criteria of the project but does not take into consideration developments that newly occurred. In short, the product might be good if made in restrictive methods, but it could have been better.

Late Testing

These methods often put the testing process late in the project. This means that the identifying and fixing of bugs is not as comprehensive as you would like them to be. After all, if everything has a set deadline and follows set protocols, you are merely finding and fixing surface-level problems, not inherent, program-breaking ones.

This is where adaptive methods are superior as the testing phase is evenly spread out across all iterations. Simply put, you are correcting your mistakes as you are building the base product.

Client Feedback Not Impactful

In most restrictive methodologies, client feedback is often ignored. And if they do acknowledge client feedback, these do not have much of an impact in the development process.

For instance, a client might want to add something to the product during the Feedback and Testing portion. Depending on how big that change is, it may be ultimately ignored so as not to change the structure of the product or haphazardly applied that it ultimately ruins the quality of the software.

High Risk

Since these methodologies are so rigid in their application, you run the risk of not addressing major problems in the coding or add enhancing features until it is too late.

Also, there is a chance that you would have to deal with constant crunch periods as deadlines for each phase are tightly set one after another. As a result, the workload of your team increases along with the pace of work. As such, you run the risk of bottlenecking your project to the point that that end product is haphazardly completed.

The Agile Process Cycle

The process of implementing the agile method differs from one strategy to another. However, they all follow roughly the same sequence, which is:

Conceptualization – Here, the product is being visualized and designed. The framework for the project will be set up and segmented, which helps in prioritizing what needs to be done. Issues like the allocation of resources and the distribution of workload will also be tackled here.

Inception – Once the project has been conceptualized, the manager must then focus on building the team (if it does not exist yet, of course). Here, the roles of each team member will be identified while the initial workloads and requirements will be designated to them.

Iteration and Construction – The most extensive part of the project, this process involves the teams going through each "sprint" or iteration as they build the product. The goal here is to present something that meets the criteria established in each iteration to upper management, shareholders, and the client.

Since the agile method is iterative by nature, it is necessary that the team goes through each of the set iterations and finish them according to the set time. At the same time, the product that they are building on must grow and develop to meet new standards and other last-minute changes per cycle.

Release – Once the base product is ready, it will undergo further Quality Assurance checks. This is where major bugs are fixed while the overall layout and user experience of the product will be revamped or enhanced.

This process will also internal and external testing, documentation of what has been fixed, and the final release of the iteration into mass production.

Production – At this phase, the developers should provide ongoing support for the software. This includes further testing and maintenance, as well as the introduction of patches to the code, if needed.

This should serve as an extra "cycle" to the process where the product is enhanced even if it has already passed the mass distribution phase. Your team can even build on the product's base features by adding more while keeping the code as functional as possible.

Retirement – Eventually, that product will reach the end of its lifespan, which lasts a year or a few after release. At this phase, the team should initiate some end of product life activities like notifying users of what is to come next and preparing them to migrate to the new product.

The sequence above presents the entire life cycle of products made using the agile model. In fact, there can be more than one agile-centric projects occurring in the same company or multiple iterations being logged in on different product lines. Better yet, the model allows a company to cater to different customers, internal or external, with their own range of needs that need to be met.

The Iteration Workflow

The agile process is dominating by cycles and iterations. Each segment of the project that is completed will actually build on the end product. In essence, with the agile method, you not only have a functional program in each iteration but also supporting features, documentation, and a code that can be used for future projects.

It is not uncommon for a project to have 3 to 10 iterations, depending on its size and type. Each iteration will also follow its own workflow, which can be visualized as follows:

Requirements – Here, the specifications of the iteration will be set. These must be based on the backlog for the product, the backlog for each cycle, and the feedback of customers and shareholders, if any.

Development – At this phase, the team develops or builds upon the software based on the goals set for that segment.

Testing – This phase will include Quality Assurance tests, internal and external training, and documentation of what has been improved or developed.

Delivery – Once the product is functional, it will then be integrated to make it cohesive. After this, the iteration of the product will then be sent for mass production.

Feedback – Once it is in the market, the development team will then monitor how the software is being received by the end-users. Are there major flaws that need addressing? What bugs did the team miss, but the customers noticed? Is there are a way to improve on the user experience? These questions can be answered at this point of the cycle.

Once the feedback phase is completed, the cycle begins anew with the team conceptualizing what needs to be done next for the new iteration. The beauty of the method is that you can come up with a better product or an entirely new offshoot in a short period of time.

What are Product Backlogs?

The most basic definition of product backlogs is that they are a list of features that can be added to an existing software created in a previous iteration. And aside from new features, backlogs can include infrastructure changes, bug fixes, and other activities that is necessary to deliver a specific outcome in a current iteration.

In other words, a product backlog answers this question:

"What can we do to make this software Better?"

Aside from the project manager, the product backlog functions as an authoritative source of what needs to be done per iteration. This means that if a task, a feature, or a fix is not on the backlog, then the development team should not even think about investing an iota of effort in performing such a task.

However, the presence of a task on a backlog does not give the assurance that the same can be delivered exactly at the end of that iteration. It only presents the team with an option on how to deliver something that was already promised at the start of the project. It is not a mandatory task that you and your team should commit to.

Perhaps your product backlog would include the following:

Increase item and weapon drops

Expand on existing world maps

Add new maps

Balance skills and classes players discovered to be over-powered

Fix game-crashing bug on Zones X3 and F10

Improve chat-based communications

Introduce Player vs. Player mode

Now, at a glance, you can determine for yourself which of the items must be added ASAP and what could be put off for the next few iterations. The point is that the backlog gives your team an idea as to what should be improved in the next iterations, so the overall product is better.

The best part about product backlogs is that you can add on them the more the product is expanded on. The addition of new features to a software gives rise to new opportunities and problems.

However, do rein in your team a bit when it comes to finishing the backlog. There is no rule that your team should clear off that backlog in each iteration. In fact, some of the items in that backlog can be introduced as entirely new features in the next project, depending on the situation.

Burndowns

Arguably, the thing that you have to deal with the most in any project is time. To be specific, you have to make sure that the progress your team is making is sufficient enough to cover the entire time period for that iteration.

And there is this fact that people outside of the development team that want you to finish your tasks *yesterday*. Their intention, after all, is always this: get things done and fast.

As such, it is the job of project managers to understand that time is an element that they must proficiently control in every project that they take. The better data they have when it comes to the time in relation to the work that needs to be done, the better a manager can make sure that their team sticks to the approved schedule.

This is where a burndown chart comes into play as it tells how much needs to be done and how much of time has been consumed by the team so far. A burndown chart is simply a graphical representation of how quickly your team is working through a customer's project.

So, how do You Read It?

Burndown charts are actually rather simple graphs. The amount of work remaining is always shown on a vertical axis while the time that has elapsed since the start and the projected end of an iteration is drawn horizontally.

The X-axis, the one that represents the timeline, is always at a straight line since the period is set. However, the y-axis representing the work that has been done or needs to be completed might fluctuate from day to day. As such, you only need to read the graph from left to right.

But, of course, the more pressing question that you might have with the chart is "what is the ideal burndown trend?" To answer that question, you have to look for certain elements in your reading.

Ideal Work Remaining – The ideal trend for this part should be a straight line connecting from the starting point to the current one. This is a telltale sign that each task has been sufficiently performed and there are no goals that have been untouched as of that iteration.

Also, at the end point, the y-axis line should cross with the x-axis. This indicates that no work is left undone.

Actual Work Remaining – But, of course, it is not exactly easy to pull off a flat line when it comes to graphs. Changes in your work plan can cause some shifts in that graph, resulting in spikes of activity in every point of the chart.

So how are you going to make this work? The best actual trend in this situation is for the actual work line to never go above the ideal work line. If the actual work line does go above the ideal, it is an indication that more work is left undone than originally planned. To put it simply, your team is way behind schedule.

But if the actual work line is below the ideal, then it tells you that your team is actually finishing their tasks on time or, better yet, you are completing the iteration way ahead of schedule.

Chapter 3: Create an Agile Environment and an Agile Team

Building an Adaptive Team

As diligent and close-to-perfect as a process can be, it is always in the service of a quality team. A good process will not magically make a bad Agile team functional. We can call any coordinated group of people working toward a shared goal a team. However, to be successful with Agile, said team must possess two specific qualities that go beyond the minimum definition: self-discipline and capability. Finding these people will make or break the entire project; beyond principles and rules, an Agile project is in the right, capable hands of the team involved. How can project managers find these critical figures?

The first step in building a team is understanding who to look for. Imagine the project manager as the head of a soon-to-be infamous heist. Maybe they need an explosives expert, a wheelman, an insider, and a hacker. Before they even approach the prospective employees, the organizer needs an exact idea of where each person goes into the project and how they will contribute to the achievement of the goal. Maybe they don't require a hacker for this project; a leader needs to understand that before they choose their team.

The same idea applies to Agile principles. When building a team, it is critical that the project manager has a clear idea of what skills, how many people, and other general boundary conditions a project may require. The importance of this analysis extends to the team once solidified. Connecting an individual to the project allows individuals within the organization to understand the relationship between roles. It further emphasizes the reason for specific drives and priorities within the project. Even if this may appear fine and dandy, there should exist a concise statement that allows all team members to connect instantly.

Creating a Vision

A compelling vision is the driving force behind successful projects. But why is it so often overlooked? Creating a compelling vision is very difficult; it requires work and leadership. Because the path forward is often intimidating and the best way not always clear, teams need something to guide them. This can, of course, be the leader who steps up in the face of intimidation, but the team must also be led on by a broader vision. There are four questions that a project manager needs to answer to develop a project vision:

1. What is the customer's product vision?

2. What are the scope and constraints of the project?

3. Who are the right participants to include in the project community?

4. How will the team deliver the product?

These questions should help define the project from a bird's-eye view. From there, a standard tool used is the simple elevator pitch. This answers a few critical questions that act as a quick reference point for anyone who might need it. The **Elevator Pitch Test** is like a game of Mad Libs:

1. **For** *[target customer]*

2. **Who** *[needs this product/opportunity]*

3. **The** *[product name]* **is a** *[product category]*

4. **That** *[key benefits, reason(s) to buy]*

5. **Unlike** *[primary alternative]*

6. **Our Product** *[statement of differentiation]*

If the Elevator Pitch Test is not the best tool for a team, there is an alternative called the **Vision Box**. A vision box is a single piece of paper that acts as a constraint. The team would then be asked to describe the project in full, based on its most critical elements in the limited space available.

Project vision should never stop being articulated as the project evolves, as the better team members understand it, the better they can make trade-offs during day-to-day operations without interference. The product vision empowers the team and directly ties to individual investment—consistency in a message is key to morale. Not only is this a tool for finding the best team, but it is also a critical aspect of maintaining the enthusiasm for a project once the team has been created. From there, everything boils down to two infamous individual characteristics:

Self-Discipline

Self-discipline is an individual journey that an individual cannot force upon themselves. Some people can absorb the mentality of self-discipline quickly, while others take longer to instill it. Self-disciplined people accept accountability, confront reality, engage, work hard, and respect their colleagues. As a project manager, understanding where each individual lands with this skill ensures better team preparation. Consequences abound for the team who doesn't possess self-discipline. Imposed discipline undermines all the core values of Agile, leading to a complete crumble of the project and team. In Kanban systems specifically, which involve zero hierarchy, self-motivation and internal drive are critical for basic functionality.

We evaluate the second paramount skill potential team members as competence. More than skill and applicable knowledge, **competence** also refers to a person's attitude and experience. Competence is the fuel behind self-discipline—it enables intense discussions, a positive and respectful approach to other people, and trust within the team. Competence also comes with a level of honesty; when an individual properly understands their strengths and weaknesses, they better understand their place on the team. However, how does anyone know their position on a team when a team can remain undefined?

Self-organizing teams are not rambunctious, lawless groups of people marauding from project to project, nor are these teams leaderless or directionless. In a self-organized team, everyone manages their own workload. From there, organizations shift work around as they need and take part in larger decision-making. The reason for the importance placed on self-discipline is precisely this: the members of the team must trust one another to complete tasks.

This does not necessarily mean that there are no natural pressures within a team. Once a team grows in experience, both with each other and with customers, the evolution leads to certain expectations. Furthermore, there are media for investigation. Each daily scrum reports exactly what each person is working on over the course of the day—those who seem to be lagging are, therefore, exposed to the whole team and often the customer as well.

Agile Leadership

These are not leaderless teams; instead, they are characterized by the shift in leadership from a manager to a coach. The shift is deliberate and difficult to achieve. An Agile project manager is in the business of guiding people rather than correcting systems. Agile leaders empower the team, encourage interactions, and facilitate decisions. At their core, leaders show a team the questions they need to answer instead of answering for them. This shift in management style is a core aspect of a team's development and sustained success—it is the piece that links the system with the people. Agile employees have defined five ways for leaders to provide better services to teams:

1. Creating a Learning Culture

Self-organizing does not implicitly mean that a team is learning and growing personally. Ultimately, investment in the organization is an investment in the company. That begins and ends with welcome training with information concisely spread across multiple days' worth of PowerPoints. Growing a learning culture within the team does not only allow the team to relax, but it also gives the highly skilled members a chance to explore something. Learning is not training; it is a chance for hands-on experimentation and working with new tools. Because the advances in modern technology are so rapid and widespread, the traditional sense of training around the emergence of new technologies is not enough. Employees need a chance to get their hands dirty with the things they want to learn, which will help them do their job better. Good leaders provide their employees with opportunities for personal growth.

2. Engaging employees in the transition

What makes a company attractive to a high-quality candidate? Is it bean bags and 401k matching? Specifically, in the modern American economy (but spreading to other corporate cultures), people are looking for opportunities in different roles. For high-quality candidates, pathways to professional and personal development are massive draws. As a company or project manager, one should rather focus on informing candidates of an impending promised change and engage them in the process. Rather than demanding employee loyalty to an idea or suggesting that movement is a recipe for failure, move employees to buy into the "why" of an organization's shift. This also requires vision from the leader, as he or she must communicate clearly.

3. Look beyond the spot market for talent

Project managers often run to HR with a list of positions, as if they were a grocery list, and ask, "How quickly can you find these people?" The HR manager will go to the broader "**Spot**" market, depending on hiring practices, and find a suitable candidate. They then process candidates, accept them, and finally welcome them to the team by having them work on a project as a new entry.

However, the perfect candidate could have been right under the management's nose the entire time. Because the hiring process is typically outwardly focused, the potential of internal hires is often overlooked. It is the job of leadership and management to make sure there are tools in place for helping internal personnel develop and making effective positions more biased in favor of experienced employees. Investments in recording and developing talents lead to happier employees—a consistent narrative.

4. Collaborate to deepen talent pool.

Aggressive efforts by companies to deepen talent pools create a tragedy of common [AC1] situations. Single companies usually swoop up as many employees as possible, but when every company does this, the result is a system that fosters a significant talent shortage. From a hiring perspective, it is in the best interest of companies to collaborate and work to understand how a few professionals can be efficiently distributed. Such an approach is not only beneficial for individual companies, but also for the "ecosystem" in which the companies co-exist and thrive.

5. Effort to managing chronic uncertainty through systematized flexibility

Modern businesses experience constant shifts in the business landscape—that fact is inescapable. However, there is a difference between externally generated uncertainty and internally developed uncertainty. The company's reaction to external uncertainty drives internal uncertainty—a rigidity within the company that requires more abrupt, dramatic shifts. This rigidity results in people losing jobs, staying where the organization doesn't need or want them, and the overall burdening of the company.

Compare two businesses looking to adopt the best in class. One offers a position in a company with compelling growth potential, whereas the second offers a similar job with similar growth potential, but also horizontal access to other business units, projects, and opportunities. It is the difference between offering a child a ladder and a sandbox. A modern company's best chance of attracting and keeping talent is to constantly look elsewhere. Offer it all and enable the employee to give as much value to the company as possible.

Programs like this are not a novel idea. Large software companies have used internal platforms for volunteering on other teams, working on other projects, and developing experience in different business units for quite a while now. Not only does this flexibility empower and support the employee as their interests develop, but it also helps the company retain talent and remain efficient through the practical application of human capital. Other resources cannot organize themselves—they require management. People are very good at determining where they want to go, though, so it is best to create systems that allow your company's employees to self-organize. This support will reward you in full.

Roles In DSDM Project Management

When it comes to applying a DSDM project management methodology, there are many different roles that need to be played and filled up by different but capable people. These roles can be grouped by interest and by actual responsibilities or functions. The following are the various interest-based roles in DSDM:

- Business-oriented roles, i.e., business perspective or expertise;

- Technical or solution-oriented roles, i.e., technical perspective or expertise;

- Leadership or management-oriented roles, i.e., leadership and general management skills or perspective; and

- Process-oriented roles, i.e., process definition and monitoring perspective or expertise.

13 Roles Team Members of A DSDM Project Need to Fill:

1. Business Sponsor: At the project level, this is the highest role or position. Because of their business-focused interest, their commitment to

63

any project, proposed solutions, and means by which to achieving them are unquestionable. They are accountable for both the project's budget and business case.

2. Business Visionary: As the name suggests, this role is responsible for giving the project its vision, identifying the project sponsor's needs, identifying the end-users of a solution being developed through the project, communicating such information to all team members, and giving the team instructions to follow for successfully completing a project. And to avoid role inconsistencies or duplications, it's best to assign this role to just one person.

3. Technical Coordinator: This role is primarily responsible for coming up with solutions that are congruent with the clients' needs, ensuring that people that fill up technical roles are adequately skilled, and making sure that the technical people are working properly. This is the technical equivalent of the previous role, the Business Visionary.

4. Project Manager: Aside from giving leadership to the DSDM project team in accordance with Agile

PMS principles, people who fill up this role are primarily responsible for managing the solution development team's working environment. Being the coordinator and facilitator of a project, they are responsible for assigning the scale and details of problems that may potentially arise in their team, which may be beyond the decision-making authority of the team.

5. Business Analyst: People who fill up this role are the only ones with multiple interests, i.e., they have sufficient knowledge of both the business needs of the project and the technical solutions that can meet the end-users' functional and non-functional requirements. Typically, business analysts belong to a Solutions Development group and are deployed at the level of projects.

6. Team Leader: Those who occupy this role have the responsibility of making development teams optimally productive, functioning as the service leader. It's somewhat akin to a Scrum Master role, in that people occupying this role should be preferably selected by work colleagues within the respective Solutions Development Teams as well as be the team's meetings facilitators.

7. Business Ambassador: People who fill up this role act as the development teams' key business representatives, particularly in the areas of prioritization and creation of requirements during the foundations and feasibility stages. Business ambassadors are responsible for every detail and prioritization in the development process. The role of business ambassador is akin to a product owner's in that it requires making business decisions for development teams.

8. Solutions Developer: People occupying this role are competent and able to improve or expand solutions by identifying and converting requirements effectively in order to ensure that all of the client's business and technical needs are met. Solutions developers function similarly to a Scrum team member, but they have skills that are more concentrated on solutions or software development.

9. Solutions Tester: Because the DSDM methodology highlights clear definition and assurance of a specific level of quality, people occupying this role are responsible not just for identifying but also for conducting tests in

accordance with an agreed upon strategy. Like the solutions developer, the solutions tester role is akin to that of a Scrum team members except that this role requires skills that are more focused on solutions testing.

10. Business Advisor: People who fill this role provide support to the team by way of business specific expertise and knowledge that other project or development team members do not possess. Business advisors are considered as subject matter experts on the business side of projects and may represent compliance or legal aspects that need to be taken into consideration, focus groups, or end-users.

11. Technical Advisor: People filling up this role have responsibilities similar to those of business advisors except that such responsibilities are geared more towards the technical areas of solutions development. Examples of such responsibilities include having extensive know-how of the technology utilized in the project, providing technical support, and identification and meeting of production or development requirements, among others.

12. Workshop Facilitator: DSDM recommends several key practices, which include workshops. People who fill the role of workshop facilitators need to be neutral, i.e., taking neither the side of end-users nor the development or project teams, and must have the ability to facilitate workshops among people of different backgrounds and attitudes very well.

13. The Coach (DSDM Coach): Adopting specific processes and mindsets necessary for transitioning into an Agile framework can be very challenging, the latter being the most challenging. In general, people who fill up this role are responsible for helping project teams deal with these challenges in order to make the successful transition to an Agile framework. Basically, think of what NBA coaches Steve Kerr of Golden State Warrior dynasty and Brad Stevens of the young and overachieving Boston Celtics do, except think of it within an Agile PMS context.

Chapter 4: What Is Six Sigma, A General Overview And A Little Bit Of History About It

A frequent buzzword, or words, Six Sigma, is well-known throughout many industries. It's a highly prized certification that immediately garners respect from top-level management. So, what exactly is it? Depending on whom you ask, Six Sigma is a program, a metric, a methodology, a system, a concept, and a management tool. Any single answer isn't quite right, although all are in some capacity correct.

Six Sigma is a management system, which relies on a few core concepts and a proven methodology. Within the Six Sigma program, the people involved will gain an understanding of the purpose and use of many management tools and cultivate out-of-the-box approaches to problem-solving. Finally, Six Sigma refers to a metric, which is how everything related to Six Sigma started.

Sigma, from the Greek alphabet, makes frequent appearances in statistics calculations and mathematics. Why? When in the upper case form: Σ, the sigma symbol represents a summation or shorthand for a specific equation. In its lowercase form: σ the symbol represents standard deviation, which is particularly important in statistics and makes the mathematical concept applicable to daily life. The idea of mathematics plays a much more significant role in Six Sigma practices than the actual use of calculation and measuring out sums. It's particularly fun to note that 6σ represents 99.7% deviation, which is the overarching goal of Six Sigma and how the program earned its name.

Within Six Sigma, you will use particular tools in various applications such as with the summation notation of Sigma in its capitalized form. You will also rely on standard deviation as the overarching goal of Six Sigma practices. The name of the system or program, Six Sigma, refers to the standard deviation of 3.4 deviations to each million. Essentially, you're reaching for perfection. In layman terms, 3.4 deviations to each million, means that for every million products created, all except four or fewer must be unflawed.

People certified in Six Sigma approach problems with fresh eyes and a new point of view. That doesn't mean that everyone involved is on board. However, when Six Sigma practitioners come into a business, they usually have the full support of top, high, and even mid-level management. They often have the freedom to implement necessary changes and help to ensure these changes stay in place and produce the desired results. However, these people don't just waltz into a business and turn the office on its head, no. Part of the training and certification is understanding buy-in, process management, and the long-term effect of getting people to want to change.

You simply can't sum up Six Sigma in a few sentences. From a high-level view, Six Sigma works to ensure that your company is 100% successful 99.7% of the time, the result of a Six-Sigma is standard deviation equations. From a close-up view, Six Sigma sometimes creates small, but always meaningful changes to control processes, eliminate waste, instill proven management methods, and more.

Six Sigma generally promises businesses that using their certified people, tools, and implementing the suggested changes will result in:

- Reduced process variation – or greater consistency

- Improved customer satisfaction – because of greater consistency

- Reduced costs – because of reduced waste and improved satisfaction

- Increased revenue – improved customer satisfaction leads to growth and reduced cost increases margins

Other benefits of Six Sigma can include boosted employee morale and more competent or capable management teams. Six Sigma is a process that will benefit all staff over a very long course of time.

Six Sigma Success Story

To understand what Six Sigma is, and the possibilities of its impact, you must see it in motion. Throughout this book, you will see numerous success stories from Six Sigma projects. The first will hopefully hit home for many people.

The Akron-Canton Regional Foodbank regularly accepts donations and then provides the donated goods to those in need. Before 2016, before two of their staff received Six Sigma training and certification, donated goods would sit in the donation center for an average of 92 days. That is over three-months that the food was possibly going to waste, or perhaps expiring. It also meant that people in need weren't getting the help Akron-Canton was trying to provide.

After their staff received training, they began looking at the processes. Akron-Canton is a non-profit organization. They don't create products. They don't even rely on a regular shipment from a supplier. From the outside, it looked as if there were no aspects of their processes that they could control. However, they reduced their holding time from 92 days to 39 days, and the team continues to work toward the goal of reducing that to 20 days.

Within the organization, they incentivized teamwork and boosted the positive impact that volunteers felt through participation. They created clear and unwavering processes for sorting, inspecting, packing and delivering food using nothing but the Six Sigma principles, methodology, and tools.

Origins and Expectations

Where did Six Sigma start? This question is one of the most popular inquiries that newly interested people have about the program. It's often followed with what to expect during training or as a business. As you learn where Six Sigma began, it will quickly become clear what to expect, both as a trainee or as a business, looking for a Six Sigma expert.

In 1986, a board member, along with an engineer and a psychologist, within Motorola, developed a program to help eliminate waste in their supply chain. The new take on supply chain management and monitoring began Six Sigma. Motorola was on board with Bob Galvin, Mikel Harry, and Bill Smith's ideas to improve quality, reduce defects, and ultimately have happier customers. At the time, Motorola was beginning its relationship with China led by Robert Galvin, then Chairman of the Board.

They were also developing pagers and reducing their products by eliminating lesser-used lines such as car radios. For the company, this was a time of massive change, many companies could not have navigated this much change without effective control from the top down, and compliance from the bottom-up. Six Sigma has many well-founded beliefs in handling organizations during times of change.

Those who come into a company with a Six Sigma background will often incite change as a means to accomplish business goals and improve the quality of the products at hand.

From the start, Motorola acknowledged that the Six Sigma program was in development and in use. They did not set out an entire plan that would be a one-size-fits-all solution. Instead, they turned their immediate focus toward aspects that all manufacturing supply chains shared. From 1986 to 1990, the developers of Six Sigma worked and reworked the Six Sigma program to identify elements in the manufacturing aspect of business controls. Known as the Manufacturing Age for Six Sigma, the only goal was to improve quality initiatives, reduce errors, and to implement metrics for quality.

During the Manufacturing Age, Mikel Harry and Bill Smith designed the classic four-phase method system used today. The MAIC methodology is part of Six Sigma teaching at every level. Through this, Six Sigma training became the prime method for teaching business professionals to measure, analyze, improve, and control nearly every aspect of their day.

After 1990, the Financial Age began. It saw the start of Six Sigma leaders starting to look past the controls within their factories. In 1990, Motorola began to garner attention from other mega-firms or super-companies such as Toyota. Working with Texas Instruments, Motorola led its first company through learning and implementing Six Sigma methodology. This interaction would set the stage that Six Sigma would be taught and then adopted by a company, rather than an ongoing position or a consulting instance. Through the Financial Age, many companies began turning to Motorola for training in an effort to turn around their operations or to better manage their resources.

Moving into the mid-1990s and beyond are the Refinement and Adoption Age. In the mid-1990s, a critical person learned Six Sigma and became its most fervent supporter. The CEO of General Electric, Jack Welch, learned of the program and invested time as well as money to implement the tools throughout all GE divisions. GE worked with the Six Sigma leaders of the time to further develop strategies and create more succinct definitions. They build the service operations programs and tools together and expanded MAIC to DMAIC.

These tools sound foreign for now, but soon you will learn what these abbreviations mean and how they are critical components of Six Sigma.

The key takeaways from the origins of Six Sigma are that it takes a look at the business from a new angle. It took one person at the very top and one person who fully understood design and development to notice that they were missing something important. When Smith and Galvin came together to build a foundation, they weren't looking at cutting corners or evading the law. They wanted instead to see how leadership could affect production, how production could impact quality, and how quality impacted the bottom line. They saw a very straight line between warehouse manufacturing processes to the end-user. Unfortunately, even today, managers, directors, and chief executives are taught to worry about their own department. Six Sigma goes beyond that, and its progression makes that apparent. Motorola could have easily kept this methodology to themselves. Instead, they sold their Six Sigma services again and again to companies that would benefit and strive because of their teaching.

They focused on improving total quality through changing processes, culture, and leadership practices. So, what should you expect? No one should go into Six Sigma believing that it will be an easy certification that will land you a ton of jobs. You should expect to be told that your current systems, processes, and even habits are wrong. How you address people or approach change may not fit into the Six Sigma principles. Many people struggle with that as they go through the training program for any of the belt levels. After the training though, you will have a much broader perspective and with the tools you need to assess and implement necessary change. What's more, Six Sigma teachings can impact a company from any level. Yes, in some cases executive-level support is essential. However, anyone can take the basics of Six Sigma and make drastic improvements to their immediate work environment and culture.

Companies should expect changes, as well. Often businesses will bring in Six Sigma experts to implement one change and then realize that the one change won't produce the results they were expecting. Any business leader or owner should know that there are possibly a thousand ways to accomplish their goals and that Six Sigma trained staff are there to help guide the business toward higher quality, fewer defects, leaner operations, and improved culture.

Here there were many success stories mentioned as you are probably well aware of Motorola, Texas Instruments, and GE. To provide a bit more tangible evidence as to the success that companies experienced and the possible futures that any of these super-companies may have experienced, this needs further explanation.

Within seven years of creating Six Sigma, Motorola saved approximately $1.4 billion in only manufacturing costs. Those cost reductions if not put into place, could have crippled Motorola. General Electric, GE, acknowledges an annual benefit of more than $2.5 billion because of its Six Sigma tools and methodologies.

CONCLUSION

We have come to the end of this book. But that doesn't mean we have to the end of this discussion. In fact, there is still plenty more to discuss. Scrum is still an evolving entity. That means that there is always something new being developed, something new being implemented, and something new being discovered.

The fact is that Scrum is the type of discipline that requires practitioners to be on a constant learning path. What that means is that you can find ways of making things work in your own field while learning about the success, and failures, others have had.

Moreover, your lack of success in one area is hardly a waste. Quite the opposite, it is a learning experience that you can use to your benefit. In that regard, things that don't go well are a way of ensuring meaningful learning experiences. These learning experiences then become lessons learned.

Consequently, all lessons learned which you can add to your knowledge base will only serve to increase your abilities in project management. Since Scrum is an evolving discipline, you have the opportunity to contribute to its evolution with your own experience and skills. This means that you can make a meaningful contribution to Scrum as a whole and to the specific field you have implemented it in.

So, what's next?

The next step is rather straightforward. Take what you have learned from this book and put it into practice in your own line of work.

If you are an experienced practitioner, there are surely aspects of this book that you can implement or even adapt to your own needs. You can certainly find ways of making the information contained in this book work to your own particular advantage.

In addition, you can use the concepts described in this book and contrast them with your own practices. This is the type of exercise that will surely help you gain a deeper understanding of your own project management style. This will go a long way toward helping you evolve your specific management style.

If you are a newbie to Scrum, then you have everything you need to get started. Perhaps this book will help you to lay the foundation for the arguments in favor of implementing Scrum in your organization. It will hopefully give you the ammunition you need to field any objections that might come your way. Often, it is the objections which you need to address far more than actual question pertaining to the feasibility of Scrum.

Ultimately, the choice of implementing Scrum is worth your while. Implementing Scrum may require your team to get some training while taking time out from their own schedules in order to get up to speed. But the fact of the matter is that Scrum is an investment that you can make in yourself and your team's capabilities.

So, what are you waiting for?

Do take the time to go over any part of this book that you feel you need to dig deeper into. Naturally, there are some parts that are easier for you to grasp than others. Some parts are easy to understand but challenging to put into practice. So, please keep in mind that practice with Scrum, not just as a means of project management but as a philosophy in the workplace, takes time to master, but it certainly pays off in the long run.

Finally, thank you for taking the time to read this book. If you have found it to be useful and informative, then, by all means, but what you have learned in this book into practice.

You can use it as a basis for training others on your team and in your organization on the ways of Scrum. The information contained herein will help you to develop a training program that you can use to disseminate the ways of Scrum in an effective and easily digestible manner.

There are many other books on this subject available in the marketplace. As such, it can be hard to pick the right one. So, do your friends and favor and recommend this one. It is a great place to start for anyone who is new to Scrum or is looking to brush up on their skills.

Thank you once again. See you next time.

Sometime in the 1980s, one of the greatest movies about martial arts was created. *Karate Kid* came like a storm - and even well into the 2000s, televisions still broadcast the movie, with nearly the same periodic recurrency as *Home Alone*.

There is a very good reason people loved that movie (as bad as it may be): it was endearing, it was about martial arts, and it gave everyone hope.

Sometime in the 1980s, one of the greatest project management methods ever created was born. *Lean Six Sigma* came as a result between Lean and Six Sigma - and it took the world by such amazing power that even today, people keep perfecting the theory, people keep learning the rules, and people still use it to save companies thousands of dollars after thousands of dollars.

Aside from the decade they were born in, and aside from the fact that they both took the world by storm, *Karate Kid* and Lean Six Sigma have one more (very important!) thing in common: their reliance on martial arts philosophy.

Sure, *Karate Kid* is but a sketch of what it actually takes to win a martial arts competition - but even so, the endearing messages and the best moments of the entire movie are the ones connected not to actual martial arts theory, but to the more romantic aspects of fighting for your title.

In Lean Six Sigma, just like in martial arts, you start low. Like Daniel-San sweeping the windows of the car in imperfect hand collaboration, you will first find that handling all the aspects of Lean Six Sigma feels a bit overwhelming - and, at times, you might even feel as if you are writing with your left hand.

The similarities don't stop here. Just like in Lean Six Sigma, Daniel had a teacher - a mentor to show him the intricacies of martial arts. And just like companies using Lean Six Sigma, Daniel-San's teachings were all about balance and routine processes that help him find his inner core of strength.

Given the fact that even the titles in Lean Six Sigma are inspired by martial arts, the comparison between this methodology and *Karate Kid* is not far-fetched in any way.

In fact, it can be fairly assumed that if you are ever in doubt about Lean Six Sigma theory and philosophy, you can simply think of a martial arts master and think of what they would do in your given situation.

Chances are that you will find an answer that is at least close to what Lean Six Sigma would propose.

Lean Six Sigma is fascinating to people for a billion reasons - and its martial arts-based nature is one of the reasons that attract curious minds towards this project management approach. It makes sense that everyone wants to become a Bruce Lee of the project management world, right?

Lean Six Sigma is a truly amazing method to employ - as long as it suits your company, of course. As it has been shown in this book, not every business and every project is meant to be applied to a Lean Six Sigma approach. In some cases, this theory is just not suitable, and it would not bring anything valuable to the table - so if you find yourself in this situation, keep the information learned in this book for "later." The chances are that you will, sooner or later, use it to fix some sort of process error in your company.

Not only is Lean Six Sigma pretty awesome from the point of view of the symbolism it employs in its naming and methods, but it is a very modern methodology as well. Back in the 1980s, people might not have cared as much about the waste reduction from an environmental point of view - but these days, this is the main buzzword you hear everywhere. And Lean Six Sigma was *there* long before "it was cool!".

We truly hope the book at hand has opened your appetite for Lean Six Sigma and everything it comes with.

While this is not even by far everything there is to Lean Six Sigma (we could talk about it for another 100 books), we hope the book at hand has helped you understand the high-level theory around this project management and problem-solving method.

We did not aim to uncover all of the Sigma secrets (or the Lean ones, for that matter). We aimed to give you a taste of just how useful, just how interesting, and just how awesome this entire framework can be. Hopefully, you have enjoyed your journey with us.

This is the second time we are using the word "journey" - and it is a very carefully chosen one, mind you.

The first one is related to the fact that once you embrace Lean Six Sigma, you will fall in love with its intricacies, with its symbolism, with the way it can actually help businesses grow bigger and healthier in so many respects.

The second one is related to the fact that Lean Six Sigma is all about continuous improvement - and what advocate would you be if you decided one day that you cannot or simply don't want to proceed further with your improvement in the art of Lean Six Sigma?

Last, but not least, the third important reason that makes Lean Six Sigma a life-long affair is the fact that it will keep surprising you, every time you use it. Sure, the theory might not seem like much when you look at it from afar - but when you see the kind of results Lean Six Sigma can bring with it, you cannot but feel really excited and productive!

Lean Six Sigma is not about empty promises of the kind you see on teleshopping advertorials. It's not a one size fits all recipe for success. And it is most definitely not a scam.

Lean Six Sigma is a way of thinking and a way of seeing life itself. When you filter actions through processes and learn to get to the root cause of things, you will be more tolerant, you will understand people better, you will have more empathy, and you will know how to treat even the more delicate situations in a way that doesn't upset anyone.

Lean Six Sigma is a method, a strategy provider, a system. Its roots may be based at Motorola and Toyota - but the system it creates is more than suitable everywhere around the world, for businesses in multiple areas of activity and of many different natures.

Six Sigma speaks internationally. It helps people from all over the world. It pushes businesses forward and, maybe more importantly than anything else, it pushes *people* forward, helping them be better, act better, grow better.

The main goal of the book at hand was not to scare you off with the myriad of information available on the topic of Lean Six Sigma.

On the contrary, actually. As mentioned above, our main goal was that of stirring your interest in this methodology and helping you understand its basics - precisely because we know that diving head-first into the more advanced techniques would feel downright scared.

The book at hand was meant to open the world of Lean Six Sigma to you and help you see that, no matter who you are and what you work, you can always pick up this theory and embrace it from the comfortable sands of a Greek island or from the comfort of your team.

Hopefully, we have provided you with the key to a new world: one where you don't have to constantly run guesswork operations on what is going wrong in a company. One where you don't want to have the responsibility of what would normally be ten other roles in a company. One where you can find actual solutions to your problems and stop "patching" them as if they are scratches on the knee.

More than anything, we genuinely hope the book here has answered your questions on what Lean Six Sigma is, how it functions, and why it can be of the utmost importance in your future.

If you are the owner of a startup, you will find this method for process improvement to be really useful.

If you are a project manager in a large company, you will definitely find Lean Six Sigma to be beneficial too!

If you are a healthcare worker, you will find that Lean Six Sigma can help you reduce the waste in your hospital so that you can focus on what you know best: saving lives.

Lean Six Sigma can be just the framework you are looking for, no matter who you may be and where you may work.

Therefore, we truly hope this book has helped you shed some light on the steps you should be following from here on, on the main philosophy behind Lean Six Sigma, and on the main techniques of its employees.

Last, but definitely not least, we hope this book was *fun* for you - because what would a learning process be without a bit of entertainment in it? Your future is about to become better because you will implement Lean Six Sigma - so what is there not to be happy about?

We wish you a cheerful, successful Lean Six Sigma road ahead of you. There might be bumps along the way, but trust us when we say that *it is all worth it*!

Scrum can help teams deliver great products on time if the team members, the Scrum master, and the product owner already have the right skills and abilities to create the product. Scrum is not a magical set of rules that any organization could just follow like a cookbook recipe and expect instant results. What I've given you is a basic understanding on the essentials of Scrum and how to use it to tap into the skills of the team members, Scrum master, and product owner, turning those into powerful leverages in creating innovative products on time.

Scrum is flexible in a sense that after several projects, it can morph into a completely different framework, perhaps with more effective tools, artifacts, and roles. Nevertheless, Scrum has no marked finish line. There is no end goal, that means you can stop learning.

Being good in the implementation of Scrum is never the end goal of companies. In the same way that studying is not the end goal of students, rather, to learn more effectively, learning to be more proficient with Scrum means you'll be able to help your company reach its goal better.

Some methodologies actually have end states, which is why they have certain levels to reach. Scrum, however, does not make an assumption that there is a state wherein you can no longer transcend the current state. It assumes that you will always find new and better ways of achieving goals. After all, this is the real world, where the best does not stay the best for too long.

No Such Thing as a Perfect Start

I've probably mentioned this quite a few times in this book one way or another, but that's because a lot of people still can't quite get this concept right. A lot of people try to implement Scrum, only to delay the start of implementation because they can't seem to perfect the concepts of Scrum. Ironically, this is exactly what Scrum is against: waiting for the perfect moment before you begin.

We live in a non-ideal world. Ideal conditions only exist in abstract mathematical notations that only a few of us would live to see. Scrum allows changes to happen because of the fact that people who try to implement Scrum will inevitably make mistakes before, during, and after the development process. A team's concept of perfection before product development will inevitably be different from its concept of perfection during and after.

If you're worried that you don't have everything perfectly planned, you should stop worrying! Perfection means no longer being able to learn new things, and Scrum emphasizes the need to continuously learn and grow. In most cases, the first few sprints may be somewhat disappointing or even downright ugly, but that's all right. The important thing is that the succeeding sprints start to become better than the previous ones, and in most cases, they really will.

Get started! By starting as soon as you can, you give your company a lot of time to grow.

The first few sprints won't be perfect and neither would the sprints in the distant future, but no Scrum implementation is ever problem-proof. All companies have problems implementing Scrum. Remember that since Scrum helps companies discover hidden kinks and bottlenecks, the companies that find these problems may associate difficulties from the problems to the Scrum framework itself. This misconception is understandable because a lot of methodologies sometimes make work seem a lot easier than it is by hiding potential problems, only to let them pop-up somewhere near the end.

Scrum is a bit more thorough, letting the teams see potential problems ahead in the beginning, middle, and end of the production. The thing is, though, Scrum doesn't tell you how to solve those problems. It can only tell you so much; the Scrum master, product owner, and team member all have to work together to find a solution.

I've said before that one can change some aspects of Scrum should he find more effective solutions, but beware that a lot of people tend to change Scrum, thinking that it'd lead to more efficient operations only to find out that they were slowly reverting back to their old methodologies. If there are dysfunctional people in the organization, the introduction of such a powerful framework that exposes bottlenecks, kinks, and other problems may make them rebellious to the idea of its implementation.

There will be a lot of impediments, especially in problematic organizations, before Scrum could be implemented. It takes patience, consistency, and diligence to properly cement the foundations of Scrum into a weakly managed organization. A lot of people, even the ones with good intentions, will rebel. People naturally resist change, especially ones that force them to change their way of thinking. Help the people involved by easing them into the principles of Scrum and give them a concrete view of the goals you're trying to achieve through this change in methodology and framework. The more people understand Scrum, the less they'll resist its implementation. The less people who resist new implementations, the better your company will be at implementing Scrum.

I don't claim that this book has all the answers you'll ever need. Far from that, I encourage you to keep learning and keep asking questions. Keep challenging existing ideologies in a healthy manner. I hope that this book has given you a lot of insights and ideas about the Scrum framework to help you in your journey in creating and delivering great and innovative software. Good luck and may your visions for your company and your teams come true!

Thanks for making it through to the end of *Kanban*. Let's hope it was informative and able to provide you with all the tools you need to achieve your goals.

Your next step is to observe and plan your transformation. Stop wondering how you can become more lean, agile, and efficient. You just read all about it! Now is the time for action. Now is the time to prepare your Kanban board and visual system to make your life easier and your team happier. Now is the time to lower costs and increase production using a simple and effective method.

While you are planning, get the buy-in from your team, company, stakeholders, and even your customers. Sell them on the benefits of adopting a Kanban system, and stay close to the process, refining as needed, so it is the most efficient system for your business. Remember, the goal of this is to assist your team members in working alongside one another efficiently while also benefiting your company. Keeping your eye on this goal during each decision you make will help with all the changes and challenges.

A Kanban methodology applies to a variety of situations, despite rumors it is "outdated." As it is with new technology, do not jump onto the glossy "bandwagon." Determine the unique needs of your organization and create a way to make this basic system work for you.. The more and more you use boards, lists, and cards, the better your team will get at running an effective Kanban project and process. As they continue to feel empowered and successful, imagine the positive atmosphere and engaged work environment you will have! Success will come to you in a variety of forms thanks to you implementing this methodology in your company. Congratulations on taking this step towards a more productive future for your company!

Inefficiency affects organizations of every type and size. Even in the most positive economic climate, companies benefit from decreasing defects and avoiding wastes. When times are tougher, enhancing efficiency can often mean the difference between keeping the business's doors open and a foldup.

Initiating Kaizen-based continual business process improvement campaigns keep companies on a growth curve.

Kaizen or 'Make Better' requires a mindset shift from an 'okay process' to 'continually better process'. The improvement is made not as a chunk, but as incremental changes accumulated over time. The changes are rapidly implemented, hence Kaizen is also called 'instant revolution'.

Kaizen may be applied to manufacturing as well as service-based organizations. Implementing Kaizen awards competitive edge & consistent growth to a business. Personal Kaizen further improves employee's quality of life both at personal and professional fronts.

Now that you have come to the end of the book, I hope that you appreciate everything that Six Sigma stands for. Over the years, there have been a lot of myths and confusion regarding this particular methodology. However, one thing should be clear by now.

Six Sigma is extremely useful for any organization that wants to improve quality, reduce costs, and enhance the speed of delivery of goods and services.

You have learned the most important tools and processes that are used in Six Sigma implementation. Keep them in mind as you move on to the next phase of the journey – which should be implementing and deploying a Six Sigma project.

Remember to follow the right procedure when trying to identify a solution to defects in your business process. Use the DMAIC stages to guide you every step of the way.

Improve the organization and control over your line, process, area, department, shift or full factory starts with being clear about what you are trying to achieve.

You need good skills around you – enough strength in depth to deal with whatever is thrown at you. A team made up of skilled, experienced people who know what is expected and what to do is the first step in the journey.

Once the team has enough key skills, it will have the capacity to deal with the day to day issues and challenges that occur in factories whilst being able to take on adopting new practices such as 5S.

5S will clear the decks of clutter, make the essentials like tools easier and quicker to locate and will give the area an organized and controlled appearance and "feel".

Keeping on top of regular red-tagging, sorts and sweeps will allow you to maintain this level of organization. You can't do it only once! It must be done until it is so habit-forming that the team do it without even thinking about it. This can take some time but it will happen if you stick with it.

Visual lean techniques like shadow and line marking will reinforce the look and feel of a controlled organized workplace.

SIC and SPC are further tools to bring in place to allow the team to quickly see and understand where they are against where they need to be.

Throughout keep using a DMAIC approach to keep on track, keep communicating to your team and everyone involved. This will create a positive "feedback" loop by allowing people to see the improvements flowing through. Seeing improvements being delivered creates a feeling of positivity and spurs teams on to achieve more.

Keep moving forward, don't quit even when things don't happen as well or as fast as you want them to, and before you know it you'll be seen as a someone who can deliver improvements and change within your business.